A PHOTO GALLERY

CATERPILLAR EARTHMOVERS AT WORK

BILL ROBERTSON

Iconografix
Photo Gallery Series

Iconografix
PO Box 446
Hudson, Wisconsin 54016 USA

Library of Congress Control Number: 2004100155

ISBN 1-58388-120-4

04 05 06 07 08 09 5 4 3 2 1

Printed in China

Cover and book design by Dan Perry

Copyedited by Suzie Helberg

Cover Photo: In David and Goliath fashion, this not so tiny bulldozer tackles a gargantuan boulder along Canada's rugged eastern seaboard near Halifax, Nova Scotia around 1956-1957. Eager to show the capabilities of Caterpillar's new "King of the Crawlers," NS Tractors & Equipment (now Atlantic Tractors & Equipment) representative Harrison Fraser demonstrated the sheer might of the new D9 as he nudged the massive piece of rock aside to make way for Bicentennial Drive—a modern highway under construction to connect the city of Halifax with a new international airport under development northwest of the city.

Table of Contents

Acknowledgments

Looking back having finished this project, I now realize that this book was years in the making. Throughout those years, many pieces of heavy equipment have been manufactured, dispatched to job sites, assigned to challenging work assignments day after day and then retired. After a productive working life, the old iron was all too often doomed to a scrap heap and that's where the story ended. In many cases, the glorious working history of these machines was recorded by means of photography—pictures that had been taken, put in a drawer and forgotten until someone decided to house clean. At this stage, unfortunately, many of these pictures then met with the same fate as the old equipment they documented. Fortunately, however, some survived and these then became the memorable classics we enjoy today.

The photographs in *Caterpillar Earthmovers at Work* are true classics! Gathered together after four decades of endless searching, many are being published for the first time in an effort to share their splendor with all who enjoy seeing Caterpillar equipment at work. I expect this book will invoke fond memories among all those who read it. I personally retain countless memories of all types of Caterpillar equipment in action or just sitting idle after a hard day's work. As we acknowledge the constructive contribution of these productive yellow giants to society, we can't understate its importance to the development of our modern world throughout the last century.

The successful completion of this book is the result of beneficial assistance from many people. I am especially thankful for the encouragement and support of the following people who helped initiate the project and then prodded me along to completion.

Brandon Lewis, proprietor of Buffalo Road Imports, Buffalo, New York is the first to receive my thanks. He identified an opportunity for this book and put me in touch with the publisher, Iconografix, Inc. Without his involvement, all the pictures in this book would still be in a drawer somewhere. Special thanks are also extended to Rettie Geldart, former superintendent with Tidewater Construction Company Ltd. of New Glasgow, Nova Scotia. Through the years, Rettie took photographs on many Tidewater projects and has graciously given permission to use them in this book. My gratitude is also extended to Kathleen Andrews of Little Harbour, Nova Scotia for pictures she provided.

In addition to the above, I want to acknowledge the unfailing encouragement and support of my family who, I believe, are a bit relieved that this project is finally completed.

It has been an education and a pleasure working with the fine talent at Iconografix, Inc. Their skill and know-how brought the book to life and I respect their effective guidance, helpful advice and consistent support throughout the project.

I hope everyone enjoys this remarkable, timeless collection of historical photographs. Sit back, relax and let your mind mosey back to the good 'ole days as you turn the pages and reflect back on a wonderful era in our construction history.

Bill Robertson
Toronto, Ontario
January 2004

Foreword

The good 'ole days...

This phrase means something different for each of us. To this now-retired heavy equipment operator, the good 'ole days could well be a fond memory of the sunny afternoon he proudly posed beside his idling Caterpillar D8 bulldozer back in the summer of 1954. This dozer, the largest made in its day, was part of a fleet of well-maintained Caterpillar equipment that earthmoving contractor Emil Anderson Construction Co. Ltd. used to build a new highway near Summerland, British Columbia, Canada. Perhaps he just nudged a huge, stubborn boulder out of the way and a proud co-worker recorded the moment of triumph. Or perhaps the operator just paused for coffee, taking a short break to enjoy the beautiful vista that lay before him. After all, once his work was finished, thousands of motorists each year would enjoy the same view, although not with as much leisure as they cautiously keep both hands on the wheel to negotiate the curving, mountainous road through south-central British Columbia.

The good 'ole days... I'll bet many of us can recall when we first saw, and heard, giant yellow earthmovers carving a path into a hillside or clearing what appeared to be a mountain of dirt in front of them. Perhaps we were young and impressionable at the time, but those noisy, greasy monsters certainly left their mark on many of us as they tackled everything that stood in their way.

Mr. Jim Hewitt, President & CEO of Hewitt Equipment Limited/Atlantic Tractors & Equipment Ltd., shares his own memories as he introduces *Caterpillar Earthmovers at Work*. We invite you to enjoy your own vibrant recollections as you examine Caterpillar equipment actively involved in road-building, pioneering access trails into uncharted timber stands, stripping topsoil for an oil refinery, placing fill in a deep gorge to support a new mountain highway, sculpting a freeway interchange, laboring in jagged rock cuts, stockpiling granular material or simply resting after a challenging day's work.

Introduction

From Newfoundland & Labrador to British Columbia, from the Niagara Peninsula to Resolute Bay in the High Arctic… for more than 75 years, "CAT" machines have played a pivotal role in the shaping of Canada as we know it today.

As the entrepreneurs who built Canada's infrastructure tackled mammoth engineering undertakings, fleets of big, yellow Caterpillar equipment have been the work horses they relied upon—rugged, dependable machines that made it possible to get the job done.

The list of these undertakings is endless: the Trans-Canada Highway from St. John's, Newfoundland to Victoria, British Columbia; the St. Lawrence Seaway; the hundreds of airports spread across this vast country; numerous, gigantic hydroelectric projects from the Columbia River in British Columbia, to Churchill Falls in Labrador, to James Bay in Quebec; the Distant Early Warning (DEW) Line defence radar system in the Arctic; the Trans-Canada Pipeline; the WW II "Alcan" Highway connecting Alaska to the rest of North America… not to mention all of the huge mining projects and forestry operations in every corner of the country. In addition, Caterpillar machines are used extensively in farming and agriculture from clearing farmland to tillage and harvesting.

One cannot help but be impressed by the sight of a large fleet of Caterpillar bulldozers and scrapers, wheel loaders, hydraulic excavators and gigantic off-highway trucks coming and going as if in an orchestrated ballet, as their operators carve the landscape and shape seemingly impregnable rock faces to build a new highway or seaport.

Today we take much of our infrastructure for granted and it may be hard for many of us to remember the days when we couldn't drive across Canada on a modern highway.

I have vivid memories of travelling with my parents and two older sisters across Canada in 1956 and of the drive from Edmonton to Jasper in Alberta (as a note of interest, we were on our way to the Canadian Good Roads Association annual convention in Jasper) over the gravel surface of a new road that was then under construction. My father, Bob Hewitt, good Caterpillar man that he was, wanted to stop and take the serial number of each Caterpillar machine that we saw. Fortunately, he didn't take them all or we never would have made it to Jasper.

But who hasn't dreamed of having the chance to just climb up on one of those huge machines, to sit in the operator's seat and just imagine the power at your fingertips?

And the people who own and operate these powerful machines are a special breed. They're the real entrepreneurs and pioneers of yesterday and today. They're the risk-takers who have helped to build the prosperous Canada that we now enjoy. At the same time, these entrepreneurs all seem to have a special place in their heart for "their Caterpillar machine." Ask any one of them and they'll tell you that their favorite model is a D8 46A bulldozer or a 966C wheel loader or maybe it's a 14E motor grader.

No matter which it is, you can be sure that they have taken very good care of that machine because, after all, that's what they relied upon to get their job done and earn their living… not just a job but also a passion.

As you turn the pages of this Photo Gallery of Caterpillar machines, I'm sure that it will evoke many memories for those of you who have operated a Cat machine and many dreams for those of you who someday hope to have the chance to do so.

Jim Hewitt
President and CEO
Hewitt Equipment Limited
Atlantic Tractors & Equipment Ltd.
September 27, 2003

Chapter 1

Harvesting Our Resources

Harvesting our natural resources is big business nowadays. It was also big business half a century ago, long before the advent of today's labor saving devices. Today, much of the work in the forests and fields of Canada is achieved by machinery—from planting seed in a farmer's fertile field to de-limbing and loading the logs of our abundant forests. Brute force, and perhaps a taste of rum, accomplished these tasks in the past. But even with the hearty workforce of the mid-twentieth century, machines were still needed to complete the work at hand. Dozers punched twisting roads through the woods to access native timber stands and allow lumbermen to get to work. Crawler tractors equipped with special application tools cleared stumps and restored fire-damaged forest areas for re-planting. Still other tractors hauled plows and operated agricultural devices to reap bountiful products from the land.

In this section, you'll see a sample of Caterpillar's contribution to these important industries as their trademark yellow equipment works to carve new access routes out of first growth timber stands, clear brush, prepare landing sites for log storage, yard wood to loading points, pile surplus chips at mill sites and assist farmers in cultivating the fields. Also depicted are the various tools and production-enhancing implements that made Caterpillar a leader in the business.

Before climate-controlled cabs and high-tracks, noisy saws used to resonate through the forest as an army of lumberjacks felled trees and stripped the trunk of its limbs. Today, a simple three-inch push of a lever directs hydraulic fluid surging through resilient hoses with enough pressure to force steel shears through the trunk of a standing tree in seconds, snipping the tree from its base and de-limbing it, almost in the same operation, prior to loading it on a waiting trailer. Rather than the sounds of buzzing saws and crashing timber, as the old-timers would have heard during their tough day's work, modern operators could well be listening to the latest news or their favorite music in the comfort of their sound-proof enclosures.

Once cleared and grubbed, fertile land in low-lying areas could then be productive for agricultural use. Generally, the smaller tractors in Caterpillar's lineup—like the D2 and D4 models—were common machines on many farms across the country. Equipped with various tools and attachments, farmers soon realized the value of these powerful little tractors in assisting them with the everyday duties of cultivation, ditching, pond construction and other work.

With no other activity in site and therefore no distractions, the operator of this 2 Ton tractor focuses his attention on the work at hand. Weighing in at 4,040 lbs. and powered by a 4-cylinder Holt gas engine, this little workhorse provided farmers with three forward speeds between 2.18 and 5.23 mph. The 2 Ton was manufactured between 1924 and 1928.

Now, this is tricky business... a skillful operator guides his Caterpillar Ten past giant boulders obstructing the trail to a lumber camp, hauling behind him a wagon of tools and supplies.

Introduced in 1938, the D2 quickly became fashionable in the farming community. This particular 4U Series, owned by Spetafore Bros., utilizes a rear-mounted power take-off attachment to operate a potato digger on the family's farm in British Columbia's fertile Fraser Delta in 1948.

This D2 idles in a peaceful orchard setting in 1946, proudly posing for the picture along with two unidentified gentlemen. The D2 was a versatile tractor for orchard use because of its size and agility. Through the years, Caterpillar offered many variations of this tractor along with functional tools and production-enhancing implements that increased the D2's adaptability even further. This particular model is fitted with an orchard exhaust system that discharged gases downward, below the tractor, to keep harmful emissions from damaging trees and fruit.

An early model D4, equipped with a cable-operated LeTourneau A4 Tiltdozer, hauls logs with the assistance of a rubber-tired logging arch in 1948.

A 6U Series D4 bulldozer clears brush on Annacis Island near Vancouver, British Columbia in August 1954. If the bulk of its duty was woods-related work, the tractor was generally fitted with a special canopy to protect the operator from unwieldy tree branches.

Although difficult to see, this 6U Series D4, equipped with a U-blade, intended to increase pushing production on the chip pile. U-blades were generally reserved for the larger, more powerful tractors but lightweight chips made for easy pushing here.

This D6 8U Series dozer positions logs at a landing site in 1949 with the assistance of an adaptable logging arch. The tractor is fitted with a dealer-supplied logging canopy as well as engine guards offered by Caterpillar to protect the motor in forestry applications. The wishbone-shaped arch structure used a fairlead on top for lift and wheels for mobility. Logs were raised off the ground and hung under the fairlead. This arrangement allowed smaller tractors to move larger logs with relative ease.

A contrast in logging implements... Above: With no protective canopy, this early model Caterpillar diesel tugs an unwieldy logging arch through the bush. Below: A 74A Series D6C tractor, fully outfitted for woods work, uses a tractor-mounted arch during clearing operations for Jay-Vee Logging near Haney, British Columbia in the spring of 1965. The integral arch on the back of the machine could be raised above the winch to provide extra lift while, at the same time, guiding cable onto the winch so it didn't burn on the edge of the drum when the load pulled to one side.

This new 74A Series D6C builds a logging road in the spring of 1963 near Kaslo, British Columbia. Equipped with a hydraulically controlled angle blade and winch, this machine could push or pull any obstacles out of the way of the new road it was building. The C-Series tractor offered about 45 extra horsepower over the 9U model it replaced.

Owned by G. P. Sawczuk Logging, this 120 hp D6C levels an area for yarding timber near Beaton, British Columbia in May 1963. This tractor was equipped with an angle dozer, winch and a rollover protective structure, commonly known as a ROPS canopy.

A D6C, fitted with upper engine guards, clears away brush for Jay-Vee Logging near Haney, British Columbia in March 1965. This particular model was also equipped with a direct, rear-mounted logging arch.

Owned by G. P. Sawczuk Logging, this 120 hp D6C pioneers a logging road into mountainous terrain on Canada's west coast in May 1963. This brand new tractor is equipped with a 6A blade, ROPS canopy and protective engine guards.

This D7 is shown working on September 1, 1954. Owned by Catermole & Tretheway, this 3T dozer, equipped with a hydraulically operated angle blade and engine guards, builds an access road along Harrison Lake, British Columbia.

On June 15, 1959 a new D7 crawler shows off a custom-designed Balderson U-blade at a sawmill near Crofton, British Columbia.

This 17A Series D7 tractor, fitted with a spark arrestor and additional lighting for 'round-the-clock operation, utilized 128 hp to economically move large volumes of wood chips for this Crofton, British Columbia mill.

Shortly after its introduction, a new D7E works for Boundary Sawmills, building logging roads in British Columbia during May 1963. This tractor is equipped with a 7A blade, engine guards and a winch—standard attire for forestry-related work.

The D8 was, beyond a doubt, the workhorse of the woods... from clearing to road building to skidding logs. Above: An RD8 tractor utilizes a hydraulically controlled CARCO blade to level a log marshalling area—one of four such machines working for Bloedel, Stewart & Welch Ltd. on this operation in 1936. Below: A tireless crawler hauls timber out to market.

Far from anywhere, this diesel tractor skids logs for Elk River Logging in the late 1940s with the help of a track-mounted arch.

Stripped of merchantable timber, this site is leveled by a D8 bulldozer for a new housing development.

A D8 8R Series tractor, equipped with LeTourneau cable-controlled bulldozer assembly, builds road near Princeton, British Columbia in 1946.

An early model D8 uses a track-mounted arch to skid timber for Elk River Logging in the late 1940s.

In these classic photographs, an RD8 tractor (above) and a Diesel 75 (below), both outfitted with radiator guards, skid logs out of their native habitat. Produced in 1935 to supersede the Seventy Five model, the RD8 became the infant of the D8 family line, wielding 95 hp at the time of its introduction. One year later, output was boosted to 110 hp and it steadily increased to the present day 310 hp D8R Series II. Both operators sit comfortably under their felt hats, one keeping a hand on a steering clutch to guide his payload into position (above) and the other shifting up to a higher gear (below).

An 8R D8 crawler, equipped with LeTourneau cable blade assembly specially designed for the D8, moves material on a new section of road under construction in a southern British Columbia forest in 1946. This model was a real producer in the bush for many years.

A common sight in the woods, these RD8 crawlers skid logs in Canada's west coast forests.

A 130 hp 2U Series D8 dozer clears logs away from the route of a proposed logging road on Vancouver Island, British Columbia in 1949.

This 2U Series D8 dozes fill into a gully to reduce the grade on a section of road under construction near Princeton, British Columbia in 1946.

A 13A D8, owned by British Columbia contractor Emil Anderson Construction Co. Ltd., clears land in the Fraser Valley. The 13A, produced between 1954 and 1955, generated 150 hp. This dozer is equipped with a powerful winch and side guarding to protect engine components from damage during clearing operations.

Now here's a moment in time... Lake Logging representatives take delivery of their brand new D8 2U Series dozer at a remote timber stand on Vancouver Island on a beautiful August 1946 morning. The tractor is equipped with a 13-foot, 4-inch wide angle blade, detached and lying flat on the same rail car. Operated by a front-mounted 24-cable control, this particular blade assembly ran 40 feet of 1/2-inch cable and could be angled to 25 degrees either side.

A familiar site in the woods of North America in the late 1940s… this D8 crawler loads a LeTourneau cable-operated pull scraper to build a main haul road.

Operating for Bloedel, Stewart & Welch Ltd. near Santa River, British Columbia this D8 and 80 pull scraper team-up to construct a logging road in 1948. The 15-yard scraper is operated from the D8 with a rear mounted 25-cable control unit. This team could ditch, grade and finish a road without the assistance of any other equipment—truly a winning combination.

On this splendid mid-May afternoon in 1956 beside Buttle Lake on Vancouver Island, British Columbia this 14A Series D8, equipped with a Fleco root rake, clears away unwanted brush from a landing point for Catermole & Tretheway logging operations. The area must have been needed in a hurry, as a D6 is busy at the site as well.

In 1956, Elk Falls Pulp Mill, needing plenty of pushing power, ordered two 15A Series D8s to manage their cedar chip pile. These torque converter tractors were chosen because of their extra pushing capability and ease of handling on the pile. The U-dozers boosted production even higher. Working in tandem, these dozers really earned money for the mill. The 191 hp 15A weighed in at 39,925 lbs. and listed at $27,200 when introduced in 1955.

In June 1965 this D8H clears land at Kelsey Bay on Vancouver Island, British Columbia. This machine was equipped with an 8A blade that was two feet wider than the 8A dozer used on the 14A model ten years before. It also sported a ROPS canopy to protect the operator in the event of a machine rollover.

Recently introduced into forestry applications just a year before, this giant D9 uses brute force to lift huge logs out of the way alongside Buttle Lake on Vancouver Island, British Columbia in 1956. This dozer, owned by Catermole & Tretheway, was fitted with a cable-controlled 9A blade and winch. With such a beautiful vista as a backdrop, it must have been hard for the operator to focus on his daily tasks.

This D9 piles blending sand high into the sky, stockpiling the material for later use on haul road structures such as culverts and bridges.

This new D9, equipped with a clearing blade, Hyster winch and full engine protection, was just moved to the site of its first job—clearing a wooded area for a new residential subdivision in North Vancouver, British Columbia in 1957.

Another view of the D9 in action, grubbing an area soon to be occupied with new homes. The cable-operated root rake quickly separated the embedded roots from soil and stone, allowing them to burn faster on the pile.

In May 1956 this Comox Logging D9 dozer clears brush near Nanaimo Lakes, British Columbia. The D9 quickly gained a reputation along the Pacific Coast as a no-nonsense producer that could easily handle large timber. It didn't take long for sunlight to flood a once-shaded wooded area once the D9 got to work.

This D9 is busy clearing land on the North Shore Mountains above Burrard Inlet, British Columbia in late January 1956.

One year after its introduction into the woods of British Columbia, the D9 was well accepted in the logging industry. The powerful D9 could successfully pioneer access roads into rugged terrain with relative ease, working on inclines up to 40 degrees.

This D9 bulldozer, owned by M. Germyn, is busy clearing land on the north shore of Burrard Inlet near Vancouver, British Columbia in late January 1956.

In May 1956 this D9, owned by Comox Logging, builds access roads through the densely forested coastal mountains of British Columbia. Owners reported that the D9 could move dirt faster and cheaper than any other machine on the market.

One west coast logging contractor claimed, in an advertisement that ran in industry periodicals in 1955, that building logging road in British Columbia is just about as rough as it gets, but the D9 was the answer to their problems. "Working eight hours a day," the ad went on to say, "the D9 with a 9A bulldozer built up to 1,500 feet of road a day, including grubbing, ditching, grading and preparing road bed for gravel." Another contractor further claimed that the D9, with 320 hp and 35 tons of weight, had the power to clear the right-of-way of everything movable!

Sold and serviced by Caterpillar Dealer NC Machinery, this superbly refurbished D9 bulldozer, equipped with a hydraulically controlled 9S blade and three-shank ripper, pioneers logging road out of the rugged Pacific coastline in the early 1970s.

Chapter 2

On The Move:
Highways, Freeways & Railways

We truly are a society on the move. By road or rail, through the air or even on foot, we require a modern and ever expanding network of inter-modal transportation facilities to assist our mobility.

During the latter half of the past century, we witnessed unprecedented growth in construction and upgrading of highways, railroads, seaports and airports around the world. During this relatively small window of time, the Eisenhower Interstate System was conceived and developed into one of the top ten engineering achievements of modern time. By any and all benchmarks, it is a hallmark of quality engineering work.

Along with highways, rail, sea and air facilities play an essential, and now integrated, part of the overall transportation system. Driven by ever-increasing volumes of passengers and freight carried by domestic and international carriers, all sectors must continue to invest, improve and modernize their networks and operating equipment.

How does Caterpillar fit in? None of these facilities would be in use today if it were not for the earthmovers used to build them. Stripping rights-of-way, grubbing unsuitable material for disposal, reshaping terrain to accommodate gentle road or rail grades, leveling huge fields for aircraft to land… all of these activities were accomplished through the use of heavy equipment.

Most common among the equipment used were the dozers and scrapers that sped up production so projects could be made ready on schedule. But many other types of equipment were required as well.

On the following pages, a wide selection of earthmoving equipment is shown removing topsoil, ripping, grading, struggling with rugged rock formations, constructing embankments, stockpiling aggregate, assisting shovels in the pit and performing other tasks. Enjoy these fascinating pictures of Caterpillar equipment doing what it does best—moving the earth… and anything else in it way.

A Caterpillar D4 bulldozer strips topsoil from a right-of-way—the material being used later to groom the slopes of this access road under construction. The 6U/7U Series D4 was manufactured between 1947 and 1959.

A D4 bulldozer mounds topsoil for John Laing & Sons in August 1954.

This 9U Series D6 dozer, equipped with 6A blade and Hyster winch, levels an area for a crusher setup near Quesnel, British Columbia.

The operator of this D6 employs a 25-cable control to empty a 60 scraper on this road reconstruction project in 1956. The 7-yard scraper, carrying 11 tons of dirt, was a popular earthmoving tool and many are still in use today.

A D6C dozer builds sub-grade on this provincial highway in Nova Scotia. Built between 1963 and 1967, the 120 hp tractor offered optional power shift transmission, improved operator visibility and had an undercarriage that was comparable in size to previous D7 models.

An RD7 crawler and Athey bottom dump wagon provide emergency assistance during removal of a rock slide that closed a section of the Canadian Pacific Railway line in the Rocky Mountains near Field, British Columbia in December 1937.

This rugged D7, equipped with cable-operated LeTourneau A7 Tiltdozer, tackles the next to impossible task of pioneering a road through this fallen rock mass in British Columbia, rock that remained undisturbed for centuries prior to the arrival of this 80 hp meddler in 1946.

A robust D7 assists a shovel to load rock from this talus slope. It looks like these men really earned their money.

This D7 dozer is constructing a highway in the summer of 1946. When not pushing, the operator used a Model K30 LeTourneau rooter to loosen material for loading. These towed rippers required tractor power of between 110 and 120 hp and generally functioned with three (detachable) teeth.

D7s proved to be a versatile tool on construction sites across the country and are shown here working in Nova Scotia grubbing, digging, piling and spreading fill.

A 102 hp 17A Series D7 operates a 70 scraper on this site in 1959. Drawn scrapers were operated by winches mounted on the back of the towing tractor, in this case a 25-cable control unit. Using rule-of-thumb measures of the day, loading time for this 10-yard scraper was only one minute. At the factory, the 70 scraper was furnished with 600 feet of 1/2-inch and 21 feet of 5/8-inch cable.

This loaded 70 scraper is shown on its way to a nearby fill (above) and unloading (below). Caterpillar production data estimated the D7 and 70 combination could carry 210 cubic yards an hour on a 400 foot (one way) haul.

Two D8s use LeTourneau pans to re-align a section of highway in 1939. This earthmoving team reportedly worked 21 hours a day—hopefully with relief operators on adjacent shifts.

A 2U Series D8 assists a DW21 motor scraper to load sandy material on this section of Ontario highway using a technique known as "pump loading."

Working on Vancouver Island near Campbell River, British Columbia a D8 tractor and a 19-yard LeTourneau scraper develop a road cut.

During August 1946 these D8 and LeTourneau scraper teams construct road for Bennett & White near Parksville, British Columbia.

This recognizable photograph, portrayed on a Caterpillar-produced puzzle in the early 1950s, depicts two D8 crawlers with 80 scrapers building road. Both tractors are fitted with protective nose plates for work in the bush.

This early model D8 pushes loam into a pile, separating it from the course, granular material below.

This operator takes a break from brush clearing activities in 1954.

Here's a contractor looking for production! Taken in August 1938 this photograph shows a section of state highway under reconstruction north of Moscow, Idaho. Two young boys look on with excitement, probably never before witnessing such activity in their quiet rural setting. This team of earthmoving monsters moved 50,000 yards of dirt over the summer with recorded fuel costs for the D8s set at only nine cents an hour.

A D8 tractor loads what looks to be a 19-yard Model W LeTourneau scraper during construction of 9.5 miles of new highway over Mt. Adin, California. The contractor's men worked 14-hour shifts on this project.

In 1946, these brand new 2U Series D8 dozers sit on rail cars, all fitted with factory-supplied canopies, ready for shipment to their new owner.

This D8 loads a LeTourneau Model LP scraper with 15 yards of dirt for W. C. Arnett on British Columbia's Hope-Princeton Highway in 1946.

Here, a D8 works in May 1955 to clear the route of an access road into the site of a new Imperial Oil Terminal alongside Lougheed Highway near Coquitlam, British Columbia.

Framed by an idle shovel boom near Cameron Lake, British Columbia in 1948, this D8 gathers material from the bank to feed the shovel during cleanup operations on this section of provincial highway.

The operator of this D8, owned by Emil Anderson Construction, pauses briefly to allow a passing photographer to capture this moment in time.

A 130 hp D8 push-loads a DW21 scraper on a section of new highway under construction in Ontario.

Sandy material is always difficult to load and, in this case, required the combined pushing power of two D8s.

A D8 pulls a cable-operated ripper to loosen material for this road construction project in British Columbia.

While it's hard to tell, it appears the operator couldn't quite navigate this muddy river and had to use a tow cable to "back track."

Along the Trans-Canada Highway near Savona, British Columbia, contractor Dawson-Wade uses two 130 hp D8s to push-load a DW21 motor scraper in October 1954.

In British Columbia's rugged Fraser Canyon, a lone D8 labors to fill a formidable gully in February 1958. Emil Anderson Construction forces were building the approach to the Saddle Rock Tunnel, which was also under construction at the time.

This D8 spreads rock for Emil Anderson on June 21, 1954. The dozer, equipped with cable-operated 8A blade, works at the site of an interchange on the Trans-Canada Highway near Langford, British Columbia.

Working in dense, sandy material, a 14A Series D8 unloads a pull scraper during construction of the Deas Island Tunnel north approach on the Vancouver-Blaine Freeway in British Columbia during March 1957.

A 15A Series D8, owned by R. A. Douglas Limited of New Glasgow, Nova Scotia, levels sub-grade for the new Trenton Park Road under construction in the summer of 1964. This dozer is equipped with 8S blade, 30-cable control and a No. 8 ripper—although the ripper was not used on this job site.

Here, another 15A Series D8 push-loads a DW21 scraper on an Ontario highway in 1957.

A 14A D8 crawler, equipped with 29-cable control operating a 12-foot straight blade, stockpiles material to feed a Pioneer crusher near Campbell River, British Columbia in May 1956 for Dawson, Wade & Macco.

Two of General Construction's D8 dozers are shown working on the new Pacific Great Eastern (PGE) railway line in 1957. Above: A 14A model, equipped with 8A blade and 29-cable control unit, pioneers a steep cut along the edge of the Peace River. Below: Working near Taylor Flats, British Columbia in August 1957 another 14A cuts an embankment to clear a route for the railway line into northern British Columbia. In the background, a new multi-span bridge is also under construction.

One of General Construction's 15As loads a LeTourneau scraper, with the help of gravity, on the new PGE railway line. This particular dozer is equipped with a 29-cable control and an elevated air intake, although close examination reveals the tractor's exhaust discharging just below the pre-cleaner. It may have missed the dust but it sure seems to catch the diesel fumes.

A direct-drive D8 tractor, owned by Tidewater Construction, bulldozes aside a mass of saturated material to reach dry road grade on this section of provincial highway in Nova Scotia in 1959.

In March 1956 a D8 14A bulldozer, fully rigged with Fleco clearing rake and Ateco ripper, tackles clearing and grubbing on the Upper Levels Highway under construction along the north shore of British Columbia's Burrard Inlet. Poole Engineering was the proud owner of this brand new machine.

Here's a portion of the dozer fleet Tidewater Construction Company Co. Ltd. employed throughout Nova Scotia in the late 1950s and early-1960s. Upper Left: A 14A Series D8 complete with 8S blade, 29-cable control and front-mounted lights strips a right-of-way. Upper Right: A D9 and D8 team-up to reposition wet material clear of a road right-of-way. Lower Left: A 14A plays in a sea of mud in the highlands of Cape Breton, Nova Scotia. Lower Right: A 15A Series D8 uses a hydraulically-controlled 8S blade to level boney material.

Upper Left & Below: Contractor Peter Kiewit Sons used D8s extensively during construction of the approaches to Deas Island (later re-named Massey) Tunnel, which carried the Vancouver-Blaine Freeway under the Fraser River. These two photos show a straight-blade crawler cutting grade at one of the tunnel approaches. Well points, installed to dewater the site, are evident in the lower photo. Upper Right: Further inland near Chilliwack, British Columbia a D8 hauls a loaded scraper to the spoil pile during grubbing operations on this section of Trans-Canada Highway through the Fraser Valley.

D8s and tractor-drawn scrapers were versatile tools on selected short haul applications. Left: A D8-463 unit, owned by Premier Construction, completes a road cut on the Vancouver Island portion of the Trans-Canada Highway. Sideboards were added to increase bowl capacity. Below: Another D8 and scraper team-up to remove topsoil from Trans-Canada Highway right-of-way just east of Chilliwack, British Columbia in July 1957.

This selection of photos shows stripping operations on the Chilliwack-Cheam section of Trans-Canada Highway during the summer of 1957. Based on a list price of $17,900 FOB factory, the 463 scraper (shown below) was estimated to cost owners $1.37 per hour of operation. Under average working conditions, it was deemed fully depreciated after 12,000 hours of service.

This D8 bulldozer, owned by Peter Kiewit Sons, carves out a portion of Trans-Canada Highway through British Columbia's rugged Fraser Canyon.

By contrast, this D8 levels sand on the Deas Island Tunnel project. Although it looks like easy work, sand was quite abrasive and could quickly wear out a set of tracks on these machines.

These photos show D8s pushing and pulling to load scrapers on two highway projects in Ontario.

Here's a brand new D8H dozer ready to roll... loaded on PGE railway car #1264, this model is fully equipped with a three-shank ripper, ROPS canopy and full track roller guards.

These D8Hs, owned by John Laing & Sons, work in a remote area of Rogers Pass, British Columbia in September 1959. In addition to increased horsepower over previous 14A/15A models, the Series H crawler also boasted 4,400 lbs. more weight.

Hanging in the sky, this new 23-ton D8H tractor is being off-loaded from an ocean-going freighter in April 1959.

This could well be the same tractor on its first job... taken only two months after the above photo, this D8H is consumed in grading work on a freeway interchange near Abbotsford, British Columbia.

D8Hs in action… Upper Left: In a remote area of British Columbia, this dozer uses a hydraulic angle blade to level boney material. Upper Right: A bulldozer works for Emil Anderson in a west coast rock quarry in 1959. Below: Another Emil Anderson bulldozer tackles a rocky slope in a new quarry under development in January 1959. The Series H tractor offered contractors 18 percent more horsepower over the 14A/15A models.

Crawlers at work in Nova Scotia in the mid-1960s… Above: A D8H fitted with factory-supplied canopy straightens a section of winding rural road. Below: In another area of the province, this D8H utilizes 270 hp to move a mound of clay during construction of one of the provincial 100-Series highways.

These two photos show special application Hystaway attachments mounted on Caterpillar tractors. Above: A 1/2-yard dragline could be mounted on either a D7 or D8 tractor. The complete assembly, fitted as shown here, weighed 10,500 lbs. plus an additional 1,310 lbs. for the dragline fairlead and bucket. Below: The Hystaway 1/2-yard backhoe weighed slightly more at 11,915 lbs. as fitted on this D8 tractor. The operator could dig 15 feet below ground and, fully extended, the hoe reached 21-feet, 6-inches at a 45-degree boom angle.

Here's the giant... Caterpillar proudly announced the D9 as the "King of the Crawlers" when introduced in 1955. Above: A new tractor is secured to a railway car for shipment. Equipped with a front-mounted 30-cable control, this D9 towers over the 13A Series D8 beside it. Below: In the field, these curious buyers examine the operator's platform on a D9 equipped with full engine protection—presumably for noise reduction during this equipment demonstration in October 1959.

Here's a selection of photographs showing one of Tidewater Construction's D9s in action on various highway construction projects throughout Nova Scotia in the late 1950s.

This Tidewater D9 bulldozer demonstrates the use of its single tooth No. 9 ripper on a highway job in Nova Scotia in 1958.

In Ontario, another D9, equipped with a three-shank ripper, push-loads a DW21 motor scraper in 1959.

This dramatic shot shows a dock crane lifting a 29-ton D9 off a freighter in 1959. This tractor is equipped with a front-mounted 30-cable control winch, popular on many tractors of the period.

More of Tidewater Construction's heavy equipment in action... Above: A swarm of earthmovers works on the Trans-Canada Highway at Mt. Thom, Nova Scotia in 1965. Below: A D9 cuts grade for a section of Highway 105 through Cape Breton, Nova Scotia.

These photos show road construction activities near Prince George, British Columbia in 1957 using heavy equipment owned by Emil Anderson Construction. For stubborn material, the D9 was a required push tractor.

In Ontario, this D9 crawler shed its blade to increase maneuverability and production during stripping operations on this section of provincial highway in 1960.

An Emil Anderson D9 push-loads a DW20 motor scraper during road construction work near Prince George, British Columbia.

Operating for Tidewater Construction Co. Ltd. in Nova Scotia, this D9 push-loads a DW21 scraper on a highway project in the early 1960s. The D9 is equipped with a 9S blade, operated by the 30-cable control unit, and a single shank ripper. Production figures for this team estimated average loading time in overburden at 45 seconds.

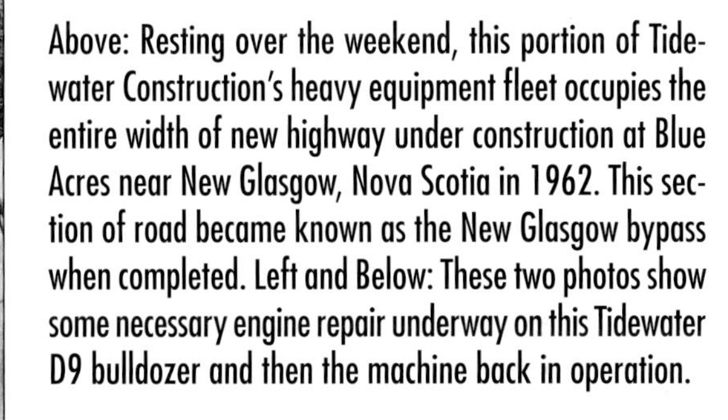

Above: Resting over the weekend, this portion of Tidewater Construction's heavy equipment fleet occupies the entire width of new highway under construction at Blue Acres near New Glasgow, Nova Scotia in 1962. This section of road became known as the New Glasgow bypass when completed. Left and Below: These two photos show some necessary engine repair underway on this Tidewater D9 bulldozer and then the machine back in operation.

A D9 push-loads a DW21 scraper on the Mt. Thom section of Trans-Canada Highway through Pictou County, Nova Scotia in the mid-1960s. This equipment belonged to general contractor Tidewater Construction Co. Ltd., then based in nearby New Glasgow, Nova Scotia.

A D9 tractor, fitted with a push-plate, loads a 20-yard DW21 scraper on this highway project in northern British Columbia.

Two 385 hp D9Gs, owned by View Construction of Kamloops, British Columbia, work on a section of Yellowhead Highway near Jasper, Alberta in September 1965.

One of View Construction's D9G dozers utilizes full horsepower to rip through hardpan on the Yellowhead Highway in 1965.

Here, one of View's 32-ton D9G bulldozers clears away rocky material as a 631B motor scraper swings around to dump its load during construction of the Yellowhead Highway in 1965.

A 66A Series power shift D9G boosts a 631 scraper out of the cut on this section of Yellowhead Highway under construction in September 1965 as another scraper enters the cut on the right.

This D9H clears organic material from an area soon to be occupied by a highway embankment in Nova Scotia.

These DW10 scrapers haul owner-supplied granular material for Campbell Contracting on the Hope-Princeton Highway through southern British Columbia in 1947. The scrapers, probably IV (tractor) and 3C (scraper) models built between 1947 and 1951, had a turning radius of 26 feet and carried just under nine yards.

A D8 bulldozer pushes a loaded scraper out of the cut as they clear away loose material on a road relocation project in Ontario.

This 115 hp DW10 tractor operates a No. 15 bowl on a road reconstruction project in Nova Scotia in 1954. The unit is owned by Chisholm Construction, Antigonish, Nova Scotia. The No. 15 bowl handled 10 yards of material per trip.

Young operators in the making…

A DW20 motor scraper dumps 18 yards of clay on this spoil pile.

Two Emil Anderson DW20s in action—these units worked long shifts to push completion of the Traverse Dam in Alberta.

Emil Anderson used the DW20 motor scraper as the major haul unit during construction of the Traverse Dam in Alberta between 1952 and 1953. Matched with a No. 20 scraper, this 225 hp unit (TP-13) hauls 18 yards of fill per trip from a nearby borrow pit for deposit on the dam.

This new DW20 motor scraper is part of a large fleet of heavy equipment that contractor Tidewater Construction had working on the New Glasgow by-pass section of Highway 104 through Nova Scotia in 1962. The company employed two DW20s along with several DW21 and 619 scrapers to bring clay material out of this cut for deposit in a low area adjacent to the nearby East River.

Two Tidewater DW20s draw clay from this cut for placement on the Trans-Canada Highway through Pictou County, Nova Scotia.

Here's a selection of DW20 scrapers... Above: Loading fill in Ontario with the assistance of a D8 pusher. Lower Left: Stripping topsoil on a British Columbia borrow pit. Lower Right: Unloading fill on a highway project near Summerland, British Columbia.

This DW21 motor scraper hauls material to the fill on a highway project in northern British Columbia. The DW21 had a turning radius of 36 feet.

Two DW21s, fitted with sideboards, load in tandem to speed up production.

Assisted by a 335 hp D9 push cat, this DW21G loads stiff clay on a road grading project near Dawson Creek, British Columbia in July 1959.

Two DW21G scrapers pass in close proximity through a highway cut in scenic Rogers Pass, British Columbia in 1959, their bowls fitted with extended side panels to increase payload per trip.

In June 1959 a 345 hp DW21 Series G tractor, matched with a 20-yard 470 scraper, works on a British Columbia road project for Prince George contractor Ginter Construction. This machine weighed approximately 30 tons (bare) and offered increased horsepower along with a 28 percent rise in torque over previous models.

Reaching the crest of the hill, a fully loaded DW21 scraper prepares to leave the cut. A Peter Kiewit Sons - Highway Construction Co. Ltd. joint venture built this section of Trans-Canada Highway through British Columbia's world-renowned Rogers Pass.

Heavy equipment strips right-of-way on the Trans-Canada Highway near Chilliwack, British Columbia in July 1957.

Here's one of the new DW21 motor scrapers Tidewater Construction had working on the New Glasgow by-pass section of Trans-Canada Highway through Nova Scotia in the early 1960s. This equipment worked on what was locally referred to as "the big cut," bringing clay out of this area to build a nearby interchange at Blue Acres.

During September 1959 these DW21 scrapers work in British Columbia's Rogers Pass, building the final section of Canada's Trans-Canada Highway. Contractor Peter Kiewit Sons used 320 hp D9s as push power to shorten the loading cycle.

During the summer of 1958, a 191 hp 14A Series D8 tractor push-loads a 15-yard DW21 scraper along the banks of British Columbia's Fraser River to cut an approach to the Deas Island Tunnel under construction in 1957.

A DW21 scraper spreads a load of sandy fill on a section of Ontario highway.

This DW21 scraper carries fill on the Vancouver-Blaine Freeway under construction in 1958. Peter Kiewit Sons owned the spread of heavy equipment building this section of roadway.

A twin-stack 21C Series DW21 scraper is pushed out of the cut by a D8 15A dozer. These torque converter tractors were preferred pushers because of their load-matching and anti-stall features. With a torque converter, the engine delivered constant horsepower output and power was automatically matched to load requirements, minimizing shifting demands on the operator.

These Peter Kiewit Sons DW21 scrapers transport sandy fill near the Deas Island Tunnel portal in 1958.

In September 1959 Peter Kiewit crews continue work on a rugged section of Trans-Canada Highway through Rogers Pass, British Columbia completing preliminary work on the right-of-way prior to moving material from side hill slopes into adjacent low areas.

Whoops… Coming back to work after a relaxing weekend away, the operator of this Peter Kiewit Sons DW21 scraper was in for a surprise. It looks like something else was occupying his seat—and luckily, he wasn't in it at the time of this mishap.

A Tidewater D9 push-loads a DW21 scraper while a D8 spreads fill on this section of Trans-Canada Highway through Mt. Thom, Nova Scotia.

This water wagon, hauled by a 300 hp DW21 two-wheeled tractor, controls dust on a road-grading project in 1960.

Miller Contracting utilizes a 619C motor scraper in July 1960 on this grading contract near Vancouver International Airport.

A D8H, equipped with a C-frame mounted push cup, loads this 619 scraper bowl to capacity.

This 300 hp, 31-ton 621 motor scraper carries a full load of fill during roadwork in Nova Scotia in the mid-1960s.

A new 621 scraper, push-loaded by a 235 hp D8H dozer, works on a new highway through Nova Scotia during the mid-1960s. This scraper could move 14 yards per trip and required a radius of approximately 40 feet to turn around — 10 feet more than the 619C model with a similar capacity.

These 21-yard, 420 hp 631B scrapers moved a lot of dirt on this project in northern British Columbia. The B Series was built in two versions—one weighing 69,700 lbs. with a 36-foot turning radius and the other weighing 70,200 lbs. with a slightly larger turning radius. Both were rated at 21 yards struck and 30 yards heaped.

This 631 uses all of its specified 37-foot, 5-inch turning radius to swing around while unloading sub-base material on this highway project near Jasper, Alberta in the fall of 1965.

D9G pushers assist 641 scrapers to load material on this highway project in western Alberta in September 1965. Judging by the size of the cut, this equipment was busy throughout the summer.

No road grade is complete until the graders move in. Here, a 112 motor grader levels a country road in November 1955.

Ready to begin a new life, this old 112 motor grader is receiving a facelift. Its owner still felt the old girl had a few good years of operating life left.

Equipment operator Dave Cook brings to life a pup motor in preparation for starting this 112 grader.

This No. 12 grader spreads gravel for Premier Construction near Duncan, British Columbia.

A 12 8T Series grader, owned by General Construction, adds the finishing touch to this section of Vancouver Island Highway.

More graders at work… Above: A 12 motor grader works on the Vancouver Island Highway in 1948. Below: In September 1959 a 115 hp No. 12 grader, owned by Peter Kiewit Sons, finishes grade on a stretch of Trans-Canada Highway through Rogers Pass, British Columbia.

The 14D grader, shown here maintaining access road near Campbell River, British Columbia in July 1963, was produced between 1961 and 1965. Rated at 150 hp, this machine weighed approximately 15 tons, required 38 feet to turn around and traveled at a top speed of 21.2 mph.

A new 14D Series grader, equipped with a Martin Grader/Scraper attachment, demonstrates its abilities in September 1963 near Matsqui, British Columbia.

Chapter 3

The Quest for Power

The quest for power—where would we be today without the essential presence of electricity? Our world revolves around it just as tiny electrons revolve around the nucleus of atoms that form the building blocks of life. From the electric toothbrush to the flick of a switch that illuminates giant sports stadiums and towering office buildings, life just wouldn't be the same without it.

Where does this power come from? Well, it is generated from many different sources like nature-fed hydroelectric stations, coal and oil-fired thermal plants, nuclear generating stations and other sources such as wind farms, tidal power stations, etc. But before energy can flow through transmission and distribution lines to our homes, factories, sports facilities and shopping centers, it must be produced at power plants, which generally are situated in remote locations. To create plants at these distant sites, all types of heavy equipment are required to construct access roads, clear plant sites and reservoirs, build dams and erect power lines that bring this vital product to the consumer.

Caterpillar equipment has been at the forefront of most large power developments around the world, uncovering fuel supplies for thermal plants and shaping landscapes to convert the lazy flow of a mountain river into an energizing force that powers a nation.

On some of the larger hydroelectric developments, for example, massive equipment must literally move mountains of material and place it in precipitous ravines to impound water for energy production. Large tractors equipped with root rakes clear away trees to form a reservoir basin. Bulldozers grub dam sites and strip borrow pits that will supply sturdy fill material for dams and other structures. Caterpillar-powered cranes lift huge penstocks into place that direct water into scroll cases and erect lattice steel structures for substations that transform electricity produced at the site and send it on its way to consumption centers.

Caterpillar uses energy to produce energy and you can be sure Cat equipment played a major role in the construction of prominent power developments everywhere. Turn the pages and enjoy, knowing the light you use to view these terrific photographs may very well come from one of these half-century old plants still producing electrical energy after all these years.

In 1946, this D7 (probably a 7M or 4T model) moved loose boulders aside to ready a quarry for production. This tractor was equipped with a LeTourneau A7 Tiltdozer powered by rear-mounted winches, which required 65 feet of 1/2-inch cable to operate. A selling feature of LeTourneau's Tiltdozer system was its versatility... a few manual turns of the crank on top of the blade support arm would change the angle of tilt at either end of the blade. Material retrieved from this pit was used by Northern Construction Company on the Bridge River Power Project—at the time the largest hydroelectric undertaking in British Columbia. The accomplishments of this project served as benchmarks for years to come as the spectacular scenery, daring in engineering design, immense scale and energy output marked it as a prized project by all accounts.

This early D7 preceded the 3T Series introduced in 1954. While the cable-operated blade offered little control and limited production when working in rock, it seemed to get the job done as the operator wrestled to clear the floor of the pit prior to the next round of drilling.

The D7E, making its debut in 1961, pushed the D7s muscle to 160 hp. A decided advantage of this machine over previous models was the more powerful hydraulic blade control system. This particular model is equipped with a Parallelogram ripper. Caterpillar literature bragged that this specific ripping concept could maintain a constant angle from initial penetration to full depth work. Estimating records of the day set an operator's wage at $3.00 an hour, but the remote location of this project, coupled with a strong union representation at this western Canadian hydroelectric site, no doubt increased this rate substantially.

In August 1946 this early model D8, equipped with a heavy-duty winch, leveled material at a dumpsite during construction of Powerhouse One on the Bridge River Development. While work on this powerhouse was postponed until 1946 as a result of the Depression and war years, the ever-growing power requirements of the post-war era led to an increased urgency to develop the site. Work then proceeded in haste and the first generator was installed in 1948 — remarkable progress even by today's standards. Between 1949 and 1954, three additional generators were brought on line to bring the plant's capacity to 180 MW — the largest single source of power in British Columbia at the time.

In May 1956 a 2U Series D8 clears the flood line along Buttle Lake, British Columbia for a new reservoir. This John Laing & Sons machine was equipped with a hydraulically controlled 8A blade and is shown working near Campbell River, British Columbia.

This D8 utilizes a cable-controlled angle blade to feed a waiting shovel during construction of an access road through the mountains.

It looks like tough sluggin' for this D8 and shovel as, together, they tackle a talus slope to retrieve riprap for a dam under construction at the Bridge River site. During the 1946 and 1947 construction seasons, Northern Construction Company must have employed an army of heavy-duty mechanics to keep equipment working in such rugged terrain.

Against a dramatic mountain backdrop, this D8 starts work on a cofferdam in August 1946. Knowledgeable operators used a technique whereby they rolled boulders and stone along the surface of new access roads under construction, which loosened fines in the material being pushed, allowing them to drop out and fill voids between the larger rocks. It didn't take long to form a relatively smooth running surface using this trick.

One of General Construction's D8 dozers utilizes a LeTourneau Angledozer to excavate a new diversion channel in August 1946 on the Elk River Power Project near Campbell River, British Columbia. Material like this must have been hard to move with cable blades.

This Series 14A D8 bulldozer labors in rock at the new Mission Dam site near Bridge River, British Columbia. Construction of the dam presented numerous site and technical challenges. To seek comfort after a few scares during the work, the "father of soil mechanics," Karl Terzaghi (1883-1963), was brought in to the rescue. He was quoted as saying: "Unfortunately, soils are made by nature and not by man and the products of nature are always complex... as soon as we pass from steel and concrete to earth, the omnipotence of theory ceases to exist. Natural soil is never uniform. Its properties change from point to point while our knowledge of its properties are limited to those few spots at which samples have been collected. In soil mechanics, the accuracy of computed results never exceeds that of a crude estimate...." With his extensive involvement in the project, the dam was later renamed Terzaghi Dam in honor of this legendary engineering figure.

Here is another view of the D8 in a sea of rock at the Mission Dam site. Generally, contractors fortified wear surfaces on blade corners and cutting edges with built-up welding to prolong the service life of hard hit ground-engaging tools under severe working conditions like these.

This D8 dozer works for the Northern Construction and J. W. Stewart joint venture building the Mission Dam near Bridge River, British Columbia. It wouldn't be hard to understand if these machines were depreciated at an accelerated rate based on these working conditions. Caterpillar handbooks of the late 1950s estimated hourly owning and operating costs for this model D8 at $9.59 under average working conditions. What multiplier would you use to reflect true operating costs in these conditions?

The mighty D9 was a powerful stripping machine at any site. With a 14-foot wide, 5-ton blade, this 230 hp dozer could move a lot of dirt.

This D9 stockpiles sand that will be used in the on-site production of concrete.

Sporting a state-of-the-art "Pushin' Cushion," this D9 could quickly approach a waiting scraper in the cut. This special application push plate could absorb the shock of machine-to-machine contact much more effectively, thus reducing operator and equipment fatigue and increasing overall production.

"Dateline: September 1964 — A beehive of activity is quickly changing the northern British Columbia landscape to create the ninth largest lake in the world." 385 hp D9Gs equipped with super-sized U-blades are busy relocating a mountain of granular material to a feed hopper. One of the longest temporary conveyor systems ever constructed worked around the clock to transport this material deposit to the dam site.

A record fleet of new Caterpillar equipment was purchased by the Kiewit, Dawson, Johnson joint venture established to bid the massive Peace River Hydroelectric Project. To win the job, these contractors effectively pooled three essential elements—experience, equipment and talent—to build one of the largest earth fill dams in the world. Hourly owning and operating costs for the D9G was then estimated at $18.15 per hour.

A 977H traxcavator, fitted with a side-dump bucket attachment, loads a truck in 1964 at the Peace River site. According to sales literature, this device increased a contractor's profit because it lowered maintenance costs and eliminated turning during loading operations, thus improving production. The 9-foot, 8-inch-wide, 3-yard bucket weighed 3,375 lbs.

These photographs show an early model PR21 rocker, weighing approximately 15 tons, loaded to capacity with around 20 yards of rock.

In selling this rocker, Caterpillar praised what they termed an "interchangeability" feature, which allowed contractors to match the DW21 tractor with a variety of other haul units, thus increasing the tractor's versatility and productivity on a job site.

A PR21 rocker hauls spoil material to a nearby dump.

Here's one of the powerful Cat 660 wheel tractors, fitted with an Athey PW660 trailer, used to build the WAC Bennett Dam on the Peace River in northern British Columbia in the mid-1960s.

In September 1964 swarms of Cat 660 wheel tractors, fitted with Athey PW660 trailers, haul fill to the Bennett Dam site. This dam was planned during the late 1950s and built in the mid-1960s, reaching completion in 1967. At 1.25 miles long, and measuring 600 feet high, it is one of the world's largest earth fill structures. The project promised relatively inexpensive electricity for the people and industries of British Columbia and the completed development supplies a quarter of the provincial power demand. Construction activity employed 4,800 workers at its peak and fueled the economy in the northeastern corner of the province. The Peace River, on which the Bennett Dam is built, carries a large water flow from a drainage basin the size of the Province of New Brunswick.

Chapter 4

Infrastructure for Industry

Industry is synonymous with growth as growth is synonymous with development. With development comes change—change in social patterns, change in movement patterns and, more significantly, change in land patterns—change brought about through the use of heavy equipment.

In remote areas, mineral resources are uncovered and mines developed. Unearthed product is then transported to more populated centers by road, rail, ship or pipeline for processing and refinement.

Before mammoth draglines and gigantic stripping shovels can begin operation at a mine site, the area must be stripped of overburden. The equipment selected for this operation depends, to a large degree, on the volume and type of material to be moved. Fifty years ago, tractor-drawn scrapers did the trick, loading and stockpiling earth and muck on short haul cycles as resources generally lie near the surface. As population, and therefore demand, grew and mineral riches retreated further beneath the ground, larger fleets of greater capacity scrapers pushed by more powerful crawlers were required.

Once mined, these underground riches were then shipped to processing plants. Dozers and scrapers built access routes to loading facilities and ports. Sidebooms dropped thousands of miles of pipeline into the ground to bring crude oil to refineries. Shovels and wagons moved rock and dirt back and forth—uncovering the raw materials being mined and then covering the exposed cavity with stockpiled earth and topsoil that would allow vegetation over the area to regenerate. At their destination, storage tanks were required to hold these reserves, then processing plants and warehouse buildings were needed to treat, process and store finished product for shipment to the consumer. Again, fleets of heavy equipment moved dirt to level the sites for these industrial facilities.

In almost every one of these separate, yet integrated, operations, Caterpillar equipment is present today, as it was yesterday. At some sites, formidable challenges could only be met and overcome by ingenious men using their Caterpillar "tools of the trade,"—tools like agile pipelayers, dependable crawlers moving mud, rock, and powerful scrapers, up to the massive 657 model, tackling huge earthmoving assignments with ease—all building the vital infrastructure required to locate, mine, develop and process resources for industry.

These MD7 sidebooms build pipeline for Imperial Oil near Port Moody, British Columbia in the summer of 1955. With a 4-foot overhang and 7,500 lb. counterweight, these pipelayers could lift 25 tons. Extending the overhang to 14 feet reduced lifting capacity to only 7 tons.

A 572 pipelayer lowers pipe for Craig & Ralston near Hope, British Columbia in 1957.

A 572 sideboom works for Craig & Ralston in 1957 near Hope, British Columbia. Introduced in 1957, this 128 hp pipelayer could lift 43 tons at a 4-foot overhang—almost double the capacity of the MD7.

On April 18, 1940 this Diesel Forty stockpiles sand for the City of Vancouver Parks Board.

This D8 tractor loads a LeTourneau Carryall scraper in September 1941, removing up to four feet of topsoil and gravel from gold-bearing property at this placer mining operation in Colorado.

A D8 tractor operates a 13-yard LeTourneau Carryall scraper to strip overburden from a vein of coal near Plymouth, Pennsylvania in 1941. Working a 16-hour shift, the tractor used 5 gallons of fuel to move 1,600 yards of material. Fuel costs averaged 7.5 cents per hour.

A D8 crawler, fitted with a rear-mounted LeTourneau power control unit that operated the pull scraper, strips soil at a placer mining operation at Horse Creek, California in 1940. After gravel had been processed, the dozer and scraper then leveled the discharged material and re-covered the area with stockpiled topsoil.

Another D8 tractor, this one equipped with a rear-mounted LeTourneau winch, operates a Model C-84 Carrimor scraper for Dutton Bros. at the Calgary Airport. The operator attributes the favorable performance of this scraper to ease of loading/unloading and a low center of gravity.

In August 1954 a D8 and 80 scraper remove topsoil for John Laing & Sons from an industrial site under development on Annacis Island, British Columbia.

Another Laing D8 spreads material in 1954, this one a 13A model equipped with a rear-mounted cable control unit that used 72 feet of cable to operate the blade.

Two 13A Series D8 dozers—one with a hydraulic blade and the other with a cable-operated blade—team-up to spread loam across this site in 1954.

These men earned their money and put their operating skills to the test at this Standard Oil tank farm site in Burnaby, British Columbia in 1954. This fleet of well-maintained crawlers and scrapers, owned by General Construction, certainly had its work cut out as it tackled this super-saturated overburden.

A new Series 13A D8 machine dozes a swell of mud in search of hard ground beneath at the site of a new oil refinery in Burnaby, British Columbia in 1954.

Above: The operator of this new 13A D8 tractor, less the unnecessary weight of a blade, guides a LeTourneau Model LP scraper through a sea of mud with the much-needed assistance of a push tractor. Below: Saddled with the same problematic site, a D8 2U tractor tugs what looks like an 11-yard LeTourneau YR Carryall scraper through the sloppy mess.

A close-up of one of General Construction's D8 crawlers used to strip this tank farm site in 1954. Here, the dozer lifts onto firmer ground, towing behind it a bowl of "soup" for disposal beyond the site perimeter.

A new 13A D8 dozes mud from the haul route, pushing it off to the side to improve traction during loading operations.

General Construction's heavy equipment tackled very challenging site conditions at this Standard Oil tank farm in 1954. With top rate tools and zealous ingenuity, men and equipment succeeded in overcoming the sloppy site conditions to reach solid ground and, today, almost half a century later, the facility is still in productive operation.

Left: A D8 equipped with hydraulic angle blade, levels sandy material on a building site surcharge in 1954.

Below: The operator of this 13A Series D8 attempts to load a LeTourneau cable-operated scraper with a saturated clay-loam mix.

Above: A bird's-eye view of an incredible lineup of new Caterpillar equipment destined for the Distant Early Warning (DEW) Line project. Taken in March 1955, all equipment used on the project was customized to suit military requirements. Below: Standing on the ground, a photographer captures the long lineup of D8s ready to go to work on this international security installation. Many of these tractors appear to be fitted with front-mounted winches for blade operation. Legend has it that most, and perhaps all, of this big iron never left the Arctic.

A D8H dozer, equipped with 8A blade and 29-cable control, stockpiles material on this spoil pile.

This D8H 46A Series dozer demonstrates its worth to earthmoving contractors at a dealer-sponsored field day in October 1959. The power shift tractor boasted 35 additional horsepower and half a ton of extra weight over the 36A model.

High on top of Porcupine Mountain near the Strait of Canso, Nova Scotia this D9, equipped with a cable-controlled straight blade, pushes a mound of crushed rock over the edge where, far below, it was loaded onto barges for shipment. Demand was constant for this high quality product and it was not uncommon to discover this granular material had been used on various construction projects all along the eastern seaboard—from Nova Scotia down to South America.

You know the old saying… "There are many ways to skin a cat." Well, there are also many ways to load a scraper. These pictures show how, if required, a DW21 scraper can be loaded by traxcavator or clamshell. Tidewater Construction Company owned this equipment and, judging by the circumstances depicted, conventional loading would not have been practical on these jobs.

When push comes to shove, the D9 did the job best of all.

A DW20 carries its load across this rough site in Ontario. No doubt a well-graded haul road through the site would have enhanced daily production figures.

With extended sideboards, this DW21 needed an extra push to top up its payload.

Caterpillar phased out the DW21 scraper with the introduction of the 619 model in 1959. It started off at a lower capacity (14 yards compared to the DW21's 19.5-yard payload) and also developed less horsepower. However, it was more maneuverable, needing only 30 feet to turn compared to the 36 feet required by the DW21. When it went out of production in 1965, it had matched the carrying capacity of the DW21. Above: A 619 operated by Tompkins Contracting moves fill near Fort St. John, British Columbia. Below: A gentleman stands beside this 619 owned by Miller Contracting during site grading operations in the summer of 1960.

Caterpillar introduced the 631 motor scraper in 1960 and production of this model continued until 1975. Over the years, horsepower and capacity remained constant while operating weight increased from 66,700 lbs. to 80,150 lbs.

A 631 motor scraper hauls its load to the fill at an industrial site in Ontario.

A 631B scraper is push-loaded by a D9G dozer at the Northwood Pulp Mill site in British Columbia in September 1964.

A 21-yard 631B strips overburden in the fall of 1964.

The two photos above (right) and below show the 32-yard capacity 651 scraper removing overburden from a large gravel deposit in northern British Columbia in the fall of 1964. The 651's weight increased from 96,400 lbs. to 124,200 lbs. over its production period from 1962 to 1984, while capacity remained the same at 32 yards struck and 44 yards heaped.

This huge 31G Series 657 motor scraper scoops up a 44-yard (heaped) load during stripping operations at the Northwood Pulp Mill site in Prince George, British Columbia. This particular series, produced between 1962 and 1968, was packed with power—a 450 hp tractor engine and separate 335 hp scraper engine. It weighed over 62 tons and required over 43 feet to turn around. It's no surprise that this brute was only used on larger earthmoving projects.

During the fall of 1964, this 651 two-wheel tractor, fitted with a Southwest water wagon, was used to add moisture to freshly placed material to assist in achieving desired density on critical embankment fills.

A Northwest shovel loads 17 yards of rocky material into a 619 rocker in September 1960 near Merritt, British Columbia.

These two photos show two rockers shedding their load… Above: A 619 rear dump demonstrates its versatility at a mining site near Merritt, British Columbia. Below: A 631 rocker works at Kennedy Lake, British Columbia in 1960.

A Bucyrus-Erie 88-B shovel loads a Cat 631 with Athey rocker in August 1961 near Kennedy Lake, British Columbia. The 40-ton Athey rear dumper was almost 25-feet long and reached a height of 21 feet when lifted. Optional equipment, available at the time of purchase, included body liners (recommended for high impact or abrasive material), a ducktail (to extend casting distance), body heating (to ease unloading in winter weather), sideboards (to increase capacity) and an auxiliary retarding brake (for shaky operators?).

A Bucyrus-Erie 88-B shovel loads a Cat 631 Athey rocker, one of several units used to strip overburden at this mine site near Kennedy Lake, British Columbia in August 1961.

This 933 traxcavator awaits a truck into which to dump its load.

This 955 traxcavator has no trouble lifting a full bucket of earth. Various models of the 955 track loader were produced, from the "C" Series in 1955 to the "L" Series in 1981. During this production range, horsepower almost doubled; increasing from 70 to 130 hp. Early models had a 1.5-yard bucket while those produced in the 1980s had a 2.25-yard bucket. In addition, weight increased from 10.5 tons in 1955 to 17.5 tons in 1981.

This 977 traxcavator demonstrates a side-dump bucket for WW Construction near Chilliwack, British Columbia in 1959.

A 966 loader moves material prior to installation of water lines at an industrial site. The 966, first produced in 1960 with a bucket capacity of 2.75 yards, is still a part of Cat's product line today.

A 325 hp 988 loader dumps a bucket of gravel into this waiting 631 scraper in 1964. The equipment, owned by Ginter Construction worked at the Northwood Pulp Mill site in Prince George, British Columbia.

A 325 hp Ginter Construction 988 loader empties a 6-yard load of gravel into a 631 scraper near Prince George, British Columbia. These 87A Series wheel loaders were produced from 1963 to 1976 and weighed approximately 40 tons.

A cable shovel loads a 35-ton 769 end-dump at this mining site in British Columbia. The early 99F Series 769, manufactured between 1962 and 1967, produced 400 hp and weighed approximately 28 tons.

A 769 truck dumps its load of rock over the edge of this steep embankment. The 769 was produced in three versions, A, B and C, between 1962 and 1995, with weights ranging from 28 tons to almost 35 tons.

More great titles from **Iconografix**

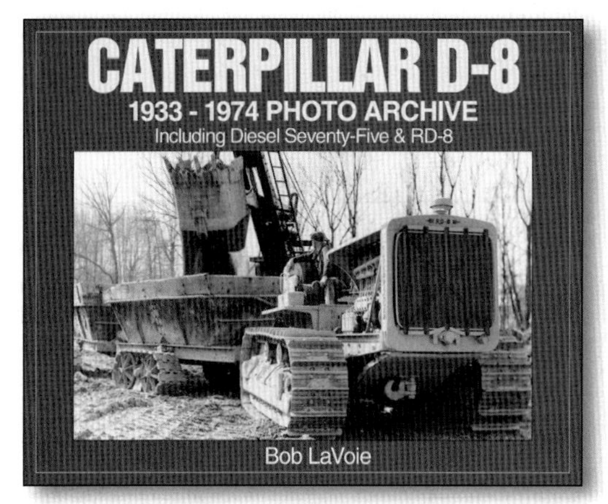

ISBN 1-882256-96-4

Call or write for a copy of our **free** catalog.

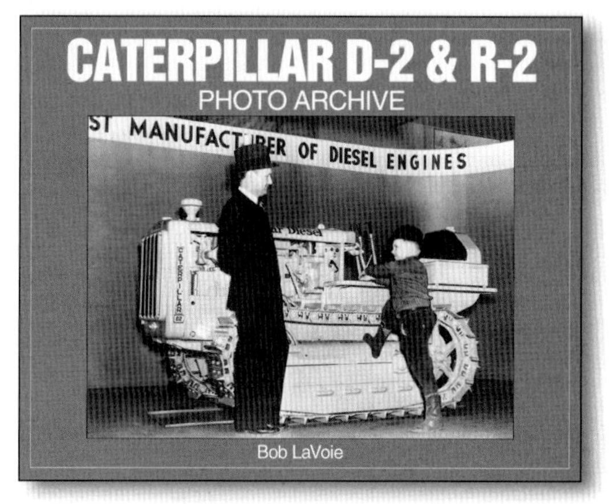

ISBN 1-882256-99-9

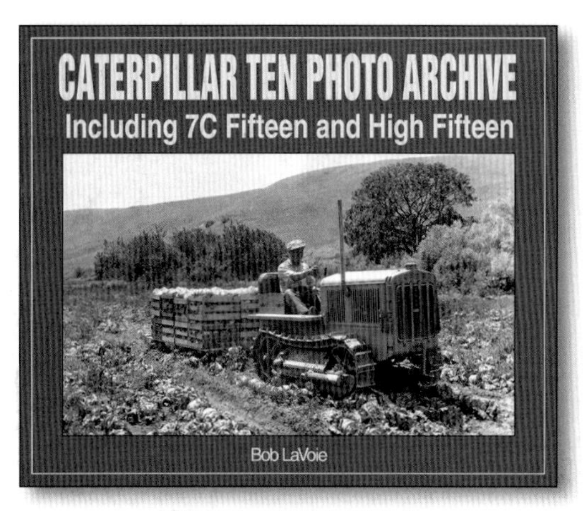

ISBN 1-58388-011-9

Iconografix, Inc.
PO Box 446, Dept BK
Hudson, WI, 54016
Telephone: (715) 381-9755,
(800) 289-3504 (USA),
Fax: (715) 381-9756

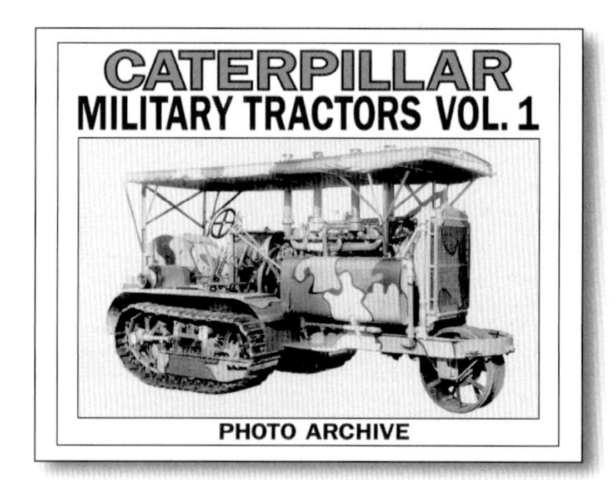

ISBN 1-882256-16-6

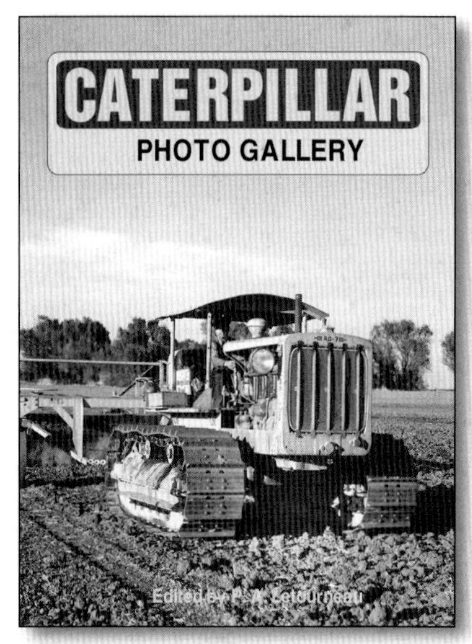

ISBN 1-882256-70-0

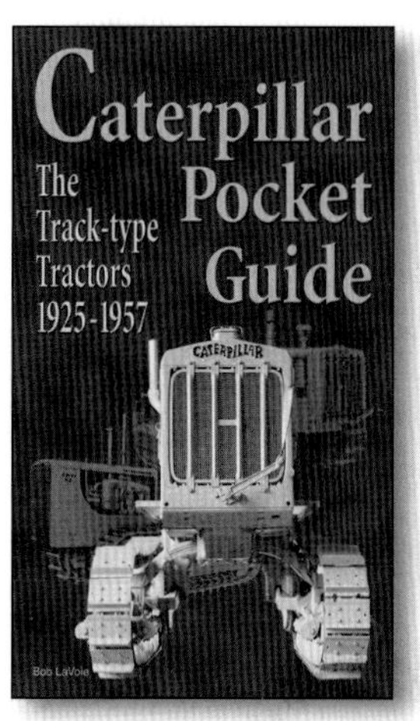

ISBN 1-58388-022-4

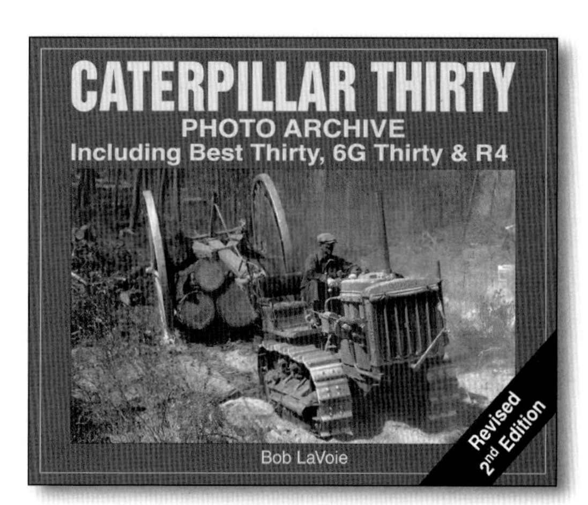

ISBN 1-58388-006-2